visit to the FIRE STATION

CHILDRENS PRESS ™
CHICAGO

by Dotti Hannum

With appreciation to the Elgin Fire Department and the fire fighters at Big Timber Fire Station, Elgin, Illinois, for their cooperation in the photographing of this book.

And to the children who worked with us so patiently as we recorded their visit to the fire station on film.

Photography by PILOT PRODUCTIONS, INC.
Dave Holmes, photographer
Jay Kelly, lighting assistant
Dean Garrison, director

Photos on pages 17, 20-21, and 23
by Sue Markson.

Photos on pages 18 and 19
by Tom Wolf.

Library of Congress Cataloging in Publication Data

Hannum, Dotti.
A visit to the fire station.

(Field trip series)
Summary: Describes a visit to a fire station and the equipment and methods used by fire fighters in their work, with tips on fire safety.
1. Fire stations—Juvenile literature. [1. Fire stations. 2. Fire extinction] I. Holmes, Dave, ill. II. Title. III. Series.
TH9148.H36 1985 628.9'25 84-12155
ISBN 0-516-01491-9

1 2 3 4 5 6 7 8 9 10 11 12 R 92 91 90 89 88 87 86 85

a visit to the FIRE STATION

Created by The Child's World

Lights flash! A siren screams! ErrrRRRrrr! ErrrRRRrrr! Quickly, the children put their hands over their ears. The lights go off. The siren stops. It's all pretend. Mrs. Dolan's class is at the fire station for a visit. They are there to see what it is like. Captain Bates will take them on a tour.

He starts by taking them into the fire fighters' classroom. "A call for help may come in while you are here," he says. "If so, the fire fighters will have to leave." He tells the children what to do if such a call comes in. Then they are ready to see the rest of the station.

Captain Bates leads them to where fire trucks are lined up, facing the door. "Do you see how the fire trucks are parked?" asks Captain Bates.

"That's so when the station gets a fire call, fire fighters can go quickly. With the doors open, the trucks can roll right out. They can rush to the fire.

"Fire fighters need special 'fireproof' clothing," he continues. "It helps keep them safe while fighting a fire."

Andy tries on a hat. Matt wears a coat. The captain shows them a pair of boots. The boots are heavy. Beth wants to try them on.

"In lots of smoke, a fire fighter wears a mask with air hooked to it," says the captain. "A mask like the one this fire fighter is holding.

"At a fire, he carries a light. Lights help the fire fighter see through the smoke.

"Next, we see a deck gun," says Captain Bates. "Most fire trucks have one. A fire fighter uses it to shoot water at a fire."

"Where does the water come from?" asks Andy.

"Some is on the truck. Or it may come from a fire hydrant," says one of the fire fighters. "It can also come from a river or pool."

Captain Bates shows the children a two-way radio. "A fire fighter uses this," he says. "With it, he can talk with someone at the main fire station or at the fire."

Then Captain Bates shows the class another truck. "This fire truck is called a hook-and-ladder truck. See the ladders," he says. "They can go up high. They are for fighting fires in tall buildings. There are small ladders too.

"Here's a picture of several firemen using a ladder to fight a fire," Captain Bates adds, as he shows the children the picture above.

"What if someone is up too high to be reached with a ladder?" Beth wants to know.

"Then that person might have to jump," says Captain Bates. "So fire fighters practice using airbags. That way they will know just how to use one when somebody needs help.

"They learn how to use a rescue net too." Then Captain Bates tells the children that fire fighters only have people jump when there is no other way to save them.

"Some fire trucks are very long," says Captain Bates. "They have two drivers. The one in front drives. The one in back just helps steer the back of the truck around corners on narrow city streets.

"Fire trucks carry many tools and things —things fire fighters use to put out fires and save people.

"Now we will go back inside.

"Here some fire fighters are checking the fire trucks," says Captain Bates. "Fire trucks are checked every morning. And they usually are washed every day."

"Where is the pole that firemen slide down?" asks Matt.

"Only older fire stations have poles," says Captain Bates. "We don't have one. No new stations do."

Next, the children look inside an ambulance. Then they go inside—a few at a time. Many fire fighters are paramedics. They know just how to help people. Things they use to help are always kept in the ambulance.

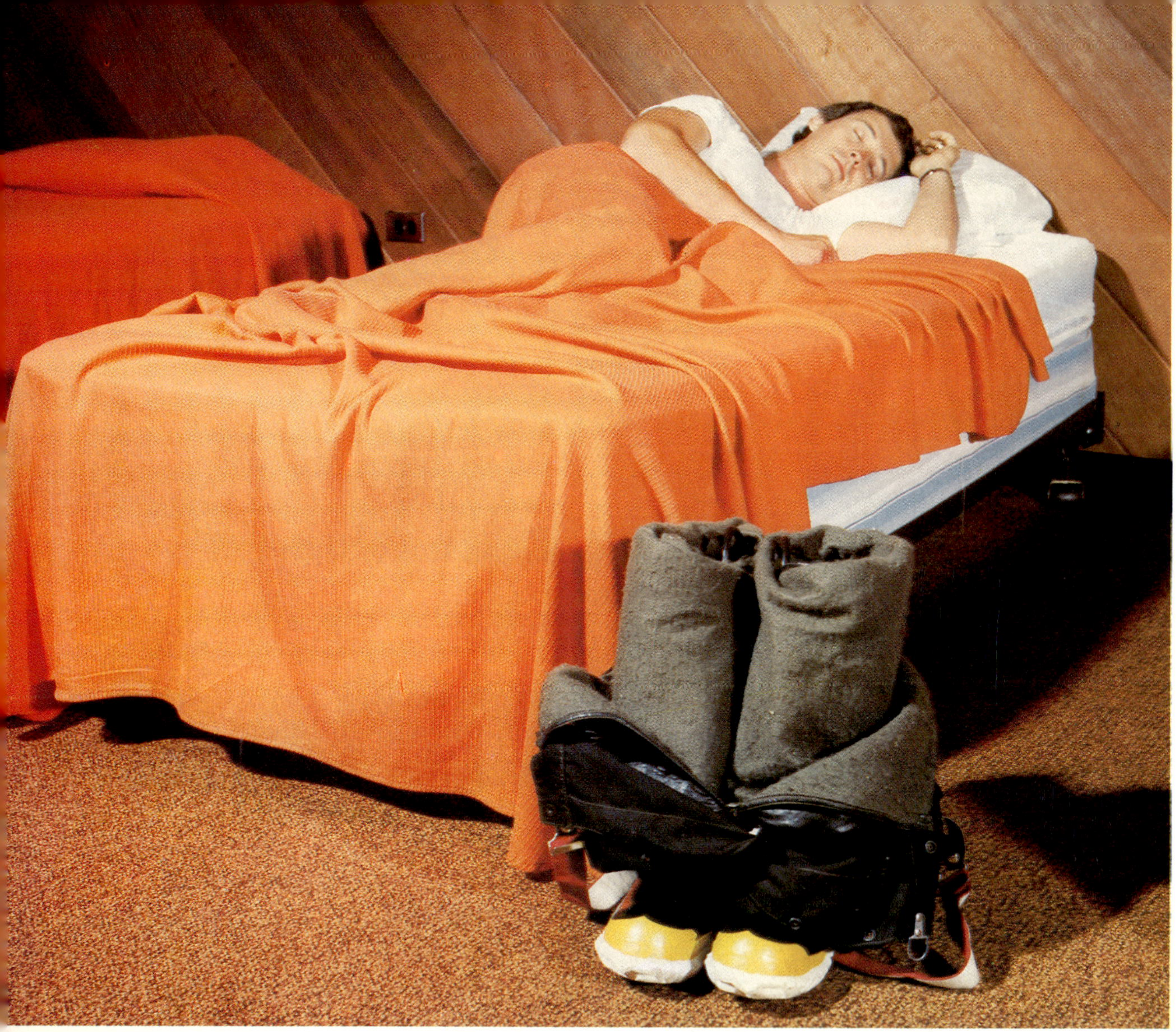

In the bunk room, Billy asks, "Is this where fire fighters sleep?"

"Yes," says Captain Bates. "Fire fighters may need to get dressed fast. So each night they hook their fire pants and boots together. And they keep them by their beds.

"Their coats, helmets, and gloves are kept on the truck where they can get them on in a hurry."

The children learn that the fire station is something like a home away from home. It even has a kitchen. "In between putting out fires, fire fighters have other work to do," Captain Bates says. "Some cook. Some shop. Others clean. Also, they make sure everything is ready to use when it's needed.

"And fire fighters have a place to exercise. They work hard at keeping in shape.

"Would you like to meet the Smoke Monster now?" he asks.

"Yes," the children say.

The Smoke Monster makes noise as he breathes.

"You can see that the Smoke Monster really is a fire fighter," says the captain. "If you ever see him at your house, you will know he is there to help you."

The Smoke Monster tells the children never to play with matches. And he says everyone should have fire drills at home.

"If you're in a building when a fire starts, you'll want to get out quickly," Captain Bates says. "But first, feel the door to the room you're in. If it's hot, DO NOT open it. Get down on the floor. Crawl to a window where you can get some air. Shout to the fire fighters. Tell them where you are.

"If your clothes or hair catch on fire, stop. Drop to the ground. Begin to roll." Captain Bates lets the children take turns pretending to be on fire. The red patch is the fire. Sue drops to roll the patch off. "You must keep rolling until all the fire is out," Captain Bates says.

At last, Mrs. Dolan's class must say goodby. "Remember that fire can be your friend," Captain Bates says. "But if it gets out of control, fire can hurt you. Then it must be put out fast. You can call us anytime. Fire fighters are always here. And we know how to put out fires fast!"